The Temples of Lower Nubia

REPORT OF THE WORK OF
THE EGYPTIAN EXPEDITION
SEASON OF 1905-'06

BY

JAMES HENRY BREASTED

ISBN: 978-1-63923-630-5

Printed: January 2023

Published and Distributed By:
Lushena Books
607 Country Club Drive, Unit E
Bensenville, IL 60106
www.lushenabks.com

ISBN: 978-1-63923-630-5

The Temples of Lower Nubia

REPORT OF THE WORK OF
THE EGYPTIAN EXPEDITION
SEASON OF 1905-'06

BY

JAMES HENRY BREASTED

THE AMERICAN JOURNAL

OF

SEMITIC LANGUAGES AND LITERATURES

(CONTINUING HEBRAICA)

Volume XXIII   OCTOBER, 1906   Number 1

Oriental Exploration Fund of the University of Chicago

FIRST PRELIMINARY REPORT OF THE EGYPTIAN EXPEDITION

By James H. Breasted

The University of Chicago

Probably there are few Egyptologists who do not realize that the monuments of Egypt still *in situ* are rapidly falling to ruin. Such catastrophes as that in the great hall of Karnak have been uncomfortable reminders of the slow but ceaseless decay which is undermining them. Griffith's timely publication of the Assiut tombs, prompted in large measure by a consciousness of this lamentable fact, was likewise an appalling witness to its truth. All who have admired in one old publication or another the transportation of the great alabaster colossus depicted on the wall of Thuthotep's tomb at el-Bersheh, perhaps do not know that this remarkable relief scene has now perished. Its gradual annihilation can be traced in the older publications as compared with that of the Egypt Exploration Fund, the last to reproduce it. Similarly, the tomb of Khamhet at Thebes, one of the most splendid published by Lepsius, has been broken out in fragments by the natives and sold to the museums of Europe. It would seem, however, that while the structural decay and barbarous demolition of the monuments are sufficiently well known, the invisible but steady disintegration of

FIG. 1.—Ramses II's Sherden (Sardinian) Bodyguard, photographed from the Wall of the Sun-Temple of Abu Simbel.

the surfaces of intact walls, especially those of the temples, involving the gradual disappearance of inscriptions and reliefs, is not generally understood. Add to this the wanton vandalism of modern visitors and native dealers, who hack out cartouches and heads, or especially well-made hieroglyphs, and the rate at which temple records are disappearing is appalling. One need only examine a series of photographs of the great geographical list of Palestinian towns recorded by Sheshonk at Karnak, and if the

FIG. 2.—Two of Ramses II's Sherden Bodyguard (same two on the right in Fig. 1), as published by Rosellini.

negatives have been made at intervals during the last twenty years, the surface of the wall from photograph to photograph may be seen slowly dissolving and the record upon it fading into blank masonry before one's eyes. Yet how few of such records exist in such a published form today that it can be called a final edition, comparable with the standard text of any important classical document of Greek or Roman antiquity! While the loss in inscribed records is irreparable, that of the reliefs is not less fatal. It seems to be the general impression that some old publication of a given relief is a final and sufficient record of the document; although it is evident at the first glance that archaeological studies based upon such a publication would be quite impossible.

Equally futile would be the study of ethnological types from such reproductions. The Sherden bodyguard of Ramses II arises before the mind's eye of every Egyptologist as a line of black-bearded, gaily costumed northerners, with spotted shields and horned helmets, crowned by enormous disks. But the beards and the spots and a good many other details, especially those of physiognomy, existed solely in the imagination of some draughtsman on Rosellini's expedition, as the reader may see in the accompanying figures (Figs. 1 and 2).

In view of the above facts, it is evident that the work of making permanent records of our fast-perishing inheritance in Egypt cannot begin too soon. The work of excavation now so ably carried on is a task of such magnitude that the organizations or expeditions now in the field can hardly be expected to undertake any additional responsibility. A grand beginning and a splendid record have already been made by the archaeological survey of the Egypt Exploration Fund, while the work of individuals, like that of Newberry, has now and then saved invaluable records. The Service des Antiquités, under Maspero's able leadership, has more than it can do with its various excavations and its never-ceasing struggle with vandalism for the preservation of the monuments. Meantime the monuments slowly perish, no precautions can save them, and the future will certainly hold us responsible for some adequate and permanent record of the documents accessible to us. Under these circumstances it seemed to the present writer the urgent duty of our expedition, for the present at least, to do what it could in this field, rather than to undertake the work of excavation exclusively.

In equipping such an expedition, the character of the records to be made conditions the methods, as well as the nature of the outfit. The reliefs and the inscribed documents of Egypt are plastic in character; any method of recording or reproducing them, such as outline-drawing, which neglects their plastic character is an insufficient record, although practical considerations, especially that of expense, may often justify such a reproduction. The loss involved, however, will be evident in such an illustration as that of the Sherden (Figs. 1 and 2), especially in the faces. Moreover, the amount of such work still undone, and the expense of doing it, demand a combination of speed, accuracy, and applicability which shall far surpass the draughtsman in these particulars. The methods at present prevailing are far too slow. I believe that the solution of the problem is to be found in the *large* camera, which, besides being far more speedy and more accurate than the draughtsman, at the same time also furnishes a record more nearly complete, in that it fully preserves the plastic character of each sculptured figure or sign on the wall. A photograph, however, represents but one illumination of the wall; whereas it may be illuminated from many different directions successively, each different illumination bringing out lines not visible before. Furthermore, the eye of the trained epigrapher, who can read the inscription and understand the broken connection, can discover more in the lacunae than the lense of the camera can ever carry to the plate. These facts demand some field process by which the speed, accuracy, and plastic character of the photograph may be combined with the reading ability, epigraphic skill, and varied illumination at the command of the Egyptologist. It was the practical solution of this problem which was attempted in the work of our expedition during the last season.

As such a method demands the development and printing of all photographs in the field, and the progress of the work involves frequent transfers from one site to another, it was found that a plate of 21×27 centimeters (about $8\frac{1}{2} \times 10\frac{1}{2}$ inches) was the largest which could be practicably handled in the narrow limits of a portable dark-room. The instrument was made to order by Kurt Bentzin, of Görlitz, and the lenses were furnished by Carl Zeiss, of Jena. It proved itself in every way equal to the demands which we made upon it. For incidental work of less importance the expedition was of course equipped with a number of smaller instruments. Our procedure was as follows: A temple wall bearing historical reliefs interspersed with the usual explanatory inscriptions, was studied as a whole, in order to determine the proper divisions into which it should be apportioned for publication. Each such division must then be photographed for each plate of the publication, but certain portions, for the sake of detail, must also be photographed again, and others still for the sake of possessing the record on a larger scale. As soon as a given section of the wall had been photographed, the negative was immediately developed on the spot. This was in order to employ the same scaffolding in making another negative, should the first prove to be unsatisfactory. If the negative was good, a print was made at once. This I took to the wall and collated from a ladder, entering on the print in red ink any readings or details which the camera had failed to record. If the photograph of any inscription proved too small for such collation, the surface of the inscription was divided into rectangles, and it was then photographed on a much larger scale. In this procedure it was a matter of indifference whether all these large-scale photographs were to be published or not; they enabled us to make an exhaustive, accurate, and rapid record of the inscription, on the basis of which it could afterward be published.

One of the problems which soon confronted us was that of lighting. It soon became evident that many portions of a temple, inside and out, are never properly lighted for a good photograph. The difficulties were of two kinds: inside walls were insufficiently lighted; outside walls received so much reflected light from so many different directions that there were no shadows, the blurred outlines disappeared, and plasticity vanished in the abundance of light. These difficulties had been foreseen, and provision made to overcome them by artificial illumination. Outside walls with too much light were then photographed at night by artificial light; insufficiently lighted inside walls were done at any time in the same way. Long study of the conditions led to the selection of pure granulated magnesium as the source of light, and a French lamp on the Nadar system proved far the best means of burning the magnesium. On windy nights out of doors, however, the Nadar lamp does not resist the draught, and it cannot be used under these conditions. We had either to wait for a calm night, or, if this was impossible, to use composition time-cartridges (*Zeitlichtpatronen*). For we never used a flashlight, but all our negatives were time exposures. Here another difficulty met us, in the fact that the expanse of wall to be illuminated was often beyond the capacity of our lamp; or, if not so, as the lamp had to be stationed at one end of the picture, that end was often overexposed, while the other end, being farther from the lamp, was underexposed (Fig. 1). We found no way out of the predicament except to treat each negative during the development, and afterward also, to correct the inequality in the negative. This method was found to be satisfactory if combined with similar manipulation of the print. Another difficulty in lighting was the inability to place the lamp in such a position that it would not shine into the lens of the camera. This was especially the case in interiors, where columns, pillars, or end walls were in the

FIG. 3.—Making a Time-Exposure in the Great Hall at Abu Simbel. The time-lamp is between the pillars on the right.

way. Very often such things forced us to place the lamp where it must inevitably shine into the lense and spoil the picture. This obliged us to erect a large black screen between the lamp and the camera, so placed that it did not intervene between the lense and the portion of wall to be photographed. This screen has sometimes to be elevated thirty feet above the floor, the lamp in such case necessarily being equally high. Under such circumstances the making of a successful exposure was a complicated matter of many hands. The photographer stood on his scaffolding with the slide of the camera drawn, and the shutter at his hand; two men with two long poles held the black screen stretched taut between the tops of the poles, while the photographer, sighting across the lense to the limit of the picture on the wall, kept them from cutting into the picture with the screen. At the same time he was also watching lest they should sway the screen too far back and expose the lense to the light. The lamp was likewise mounted on the end of a long pole, held erect by a man standing below, while another, standing at the top of a ladder beside the lamp, was ready at the word to pump a steady stream of granulated magnesium into the alcohol flame of the Nadar lamp, already lighted for the purpose. Exposures as long as twenty-five seconds were made in this way with complete success.

Space forbids any discussion of the methods which we devised in connection with the dark-room, the development of plates, and the use of the turbid Nile water in developing and washing them. I hope in some future report to return to this side of the enterprise. Occasionally it was possible to illuminate a large expanse of wall by throwing daylight upon it from large reflectors at a sharply acute angle with the wall, a comparatively small beam of light thus covering a large surface.

These reflectors, however, proved much more useful in another class of work, in which the camera could not be employed. We carried with us a quantity of new and very bright sheet-tin. With this it was possible to build large reflectors, by tacking the overlapping sheets upon a wooden frame, which could be easily carried about and propped up at any angle. Now and again an inscription was found which was too badly weathered to be photographed. In the illumination of such inscriptions these reflectors were invaluable. The great stela of Ramses II's marriage to the Hittite princess, at Abu Simbel, is for the most part so badly weathered that a photograph of much of it would be of little value. At the same time it is so situated that its long inscription becomes one blur of rough sandstone in the abundance of light which falls upon it. It was necessary to make a

FIG . 4.—Method of Sha ding and Il minating from a Re flector flecte dat OneSi de .
The sunshine on the inscription before the copyist on thescaf folding comes from the re flector
aside the native in the foregro nd . The inscription covere dwith canvas to exclu e the
xcess of light is the stela at AbuSimbel recording Ramses II'smarriage to a Hittite
fincess .

hand facsimile of this inscription. It is in forty-one lines, each about eight feet long, making about three hundred and twenty-eight feet of inscription. It was impossible to copy so long an inscription at night. It became necessary, therefore, to control the light during the day. I had a scaffold built before it, and inclosed the whole scaffold, except the lowest portion, in canvas, thus darkening the whole monument and cutting off the embarrassing quantity of light. At the left end, close to the surface of the stone, the canvas was then drawn back. At the same side, fifty feet away a reflector was set up, and through the opening at the left end of the canvas it shot a broad beam of sunshine at an acute angle upon the darkened surface of the stone. A native was placed in charge of the reflector, and he soon learned to change its position from time to time, as the sun moved. This furnished the sharply oblique light necessary to the reading of such an inscribed surface. Nevertheless no half-illegible monument can be exhausted if the light come from one direction only. It is surprising how seemingly illegible signs will suddenly come out clearly if the direction of light can be widely varied. To accomplish this the word under examination could be covered with the left hand, thus cutting off the sunshine from the left. At the same time I held at the proper angle on the right side of the word a hand-mirror, which caught the sunshine coming from the large reflector and threw it upon the word from the right. This process can probably be discerned in the cut (Fig. 4). In the same way, by cutting off the light from the left, it could be caught in the hand-mirror as it came from the large reflector, and be thrown from above or below the word as desired. This method brought out many new readings on the great stela of the Hittite marriage.

It became necessary, in undertaking such a campaign, to determine how exhaustive the record of inscriptions and reliefs should be. It is evident that all reliefs and inscriptions of *historical* import should be included. As for those of *religious* character, there are in most temples such a host of purely conventional scenes that it would be of slight use and involve a prohibitive size and price for the publication, to include all these. I refer, of course, to the countless tableaux in which the king appears offering to some divinity. In such scenes the name of the cultus act depicted, the name and titles of the divinity, and sometimes the word or two uttered by the god or the king are of

importance, but especially the first two. It was therefore determined that our survey should include a catalogue of all scenes
and inscriptions of whatever character. Of these all historical
documents, and all offering scenes of size and importance, should
be reproduced *in toto;* but the innumerable small offering and
other religious scenes which cover the columns, pillars, pilasters,
and door-thicknesses of the temple should find place only in the
catalogue, which in all cases should record the name of the cultus
act wherever given, the name and titles of the divinity, and the
speech of the god or king, wherever they were of any importance. It was decided also that, so catalogued, all of this material
should be published. It is conceivable that the costume of the
monarch and of the gods, as well as the objects offered in these
scenes might in some cases, be of some archaeological value,
although as a whole hardly commensurate with the amount of
time and expense involved.

While architecture in general was not to be wholly overlooked,
it seemed wise not to burden the enterprise with responsibility
for a detailed architectural survey; but rather to make the main
object of the expedition an *epigraphic* survey. It is to be
devoutly hoped that an *architectural* survey of the widest scope
may be undertaken among the temples of Egypt before it is too
late, but the modern requirements for such a survey can hardly
be met by an expedition undertaking also the responsibility for
a complete epigraphic survey.

Probably there is no region controlled by ancient Egypt where
so little epigraphic work has been done as in Nubia. It was
therefore decided to begin on the temples there. This field lies
partially within the jurisdiction of Egypt, partially within that of
the Sudanese government. Application was made to both these
governments for permission to begin such a survey. His excel-
lency Sir Reginald Wingate, governor-general of the Sudân, was
most cordial in his expression of interest in the undertaking, and
granted the expedition full permission to carry on the desired
work. For this very kind interest, and the generous permission
extended the expedition, I would here take public opportunity to
express the thanks of the expedition, as well as my own deep
appreciation of the frank and cordial reception accorded our enterprise. From the committee controlling such work in Egpyt also
our undertaking met with very favorable consideration, and in a

series of conferences with M. Maspero I found him ready to co-operate most willingly with our plans. I would take this opportunity also publicly to express to him my thanks for his ready assistance and cordial co-operation. After much delay in obtaining our equipment and general outfit, which were very slow in arriving, the expedition embarked at Assuan in the dahabiyeh "Abu Simbel," and on Christmas Day, 1905, began the voyage southward to begin work in the Sudân, at the lower end of the second cataract, in the two temples of Halfa. Various unforeseen circumstances had combined to make it impossible to begin at the fourth cataract, the southern limit of Egyptian conquest, and work northward. Indeed, we had been so delayed that the water would have been too low on the upper cataracts to attempt their passage. It was therefore decided to postpone Upper Nubia until the next year, and to complete the work in Lower Nubia, between the first and second cataracts, first. It was further determined to confine our efforts to the pre-Ptolemaic temples of the region in question, as the most important monuments existing there. Besides the present writer, the personnel of the expedition consisted of Mr. Victor S. Persons, whose training had been that of a civil engineer, with a season's experience on the Babylonian Expedition of the University of Chicago; and Herr Friedrich Koch, of Berlin, who served as technical photographer. We had also seven or eight native assistants and helpers, six of whom served as crew of the dahabiyeh at the same time, whenever the expedition was in motion; and besides these, three native servants.

On Sunday, January 7, 1906, the expedition arrived at Halfa, and the season's work was begun the next day. There are two temples here, on the west shore opposite the town of Wadi Halfa. It was Captain Lyons who first excavated the southernmost of the two buildings. The town is the Bohen of the early Egyptians, known to the Ptolemies as Boon. The larger temple was built by Thutmose III, and does credit to his architects. It contains some very fine reliefs, with the colors still preserved in surprising freshness. Nothing at the place has ever been published save the graffiti in the first hall, which have been partially edited by Sayce. The expedition therefore undertook a complete record of all the inscribed or sculptured materials in the building.

Apart from their artistic value, the walls of this temple exhibit graphic evidences of the feud in the royal family at Thebes. In

FIG. 5.—Temple of Thutmose III and Hatshepsut at Halfa. The roof, the square pillars bearing it, and the surrounding wall are modern works for the protection of the temple, erected by Mr. Scott-Moncrieff, of the British Museum.

the relief scenes the entire figure of Hatshepsut, the queen, has
been cut from the wall, to the depth of six inches. New blocks
were then let into the wall, where it had been so cut out, offering
a fresh surface for the insertion of the figure of Thutmose II.
Such thorough erasure and replacement has never been found
elsewhere. The references to the queen in the inscriptions have
also been carefully expunged, all feminine pronouns and endings
having been replaced. by the corresponding masculine forms, and
the name of the queen itself by that of Thutmose II. On several

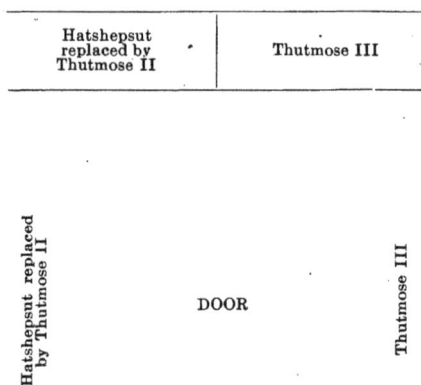

Hatshepsut replaced by Thutmose II	Thutmose III

Hatshepsut replaced by Thutmose II   DOOR   Thutmose III

FIG. 6.—Door Built during Coregency of Thutmose III (right) and Hatshepsut (left),
on which Hatshepsut has been supplanted by Thutmose II.

of the doors such substitution is very significant. On one door-
post are the names and titles of Thutmose III, and on the other
originally the name and titles of Queen Hatshepsut, showing
clearly that the door was erected during the coregency of Thutmose III and his gifted queen, Hatshepsut. But Hatshepsut's
name and titles have in every case been replaced by those of Thutmose II, showing that he must have interrupted the coregency of
Thutmose III and the queen.

In the rearmost room of the temple, however, the ·figure of
Thutmose II occurs once in such a way as to show that both it
and the accompanying name and titles are original. It would
thus appear that the reliefs and inscriptions of the temple were

not entirely completed at the time when the name and figure of Hatshepsut were superseded by those of Thutmose II. His artists therefore found a still vacant wall in the last room where they could insert his name and figure, without superimposing them upon those of the queen. The explanation of the traditional school that these insertions of Thutmose II's name over that of Hatshepsut were made by Ramses II, who has not a single document in the temple, is quite impossible here. The

FIG. 7.—Middle Kingdom Temple at Halfa.

insertions can be due only to Thutmose II himself, and this throws his brief reign after at least a portion of the coregency of Thutmose III and Hatshepsut. Any unprejudiced observer, not knowing the order of succession of these rulers, could only conclude from the phenomena exhibited by these walls that, after Thutmose III and his queen Hatshepsut had erected this temple, he whom we know as Thutmose II expunged the queen's name for some reason, and inserted over it his own.

The neighboring Middle Kingdom temple, in which Champollion found the stela of Sesostris I, contains few inscriptions, but I found there a graffito of the viceroy of Kush, Paynehsi, heretofore known to us only through Ramses XII's letter to him (see *Ancient Records*, IV, §595). No records of the Middle Kingdom Pharaohs survive in this temple at the present day. Of the Empire rulers I found only Amenhotep II and Seti I. Those portions of the building surviving seem to date from the Empire, and I take it that the stela of Sesostris I, which was found by Champollion in the Holy of Holies, was re-erected there by some Empire ruler, as the stelae of Sesostris III at Semneh were preserved and set up in the Empire temple at that place by Thutmose III.

The ancient town was south of the temples. On the north-

FIG. 8.—Remains of Ramses II's Temple at Aksheh.

west I found a graffito on the rocks, and below these the native sailor who was with me found relics of more modern history. They were the fragments of a shrapnel case, together with some of the bullets scattered by its explosion. It had burst there in some action between the British and the dervishes. An isolated peak of rock west of the town bears numerous names of workmen, who were evidently employed here not long after the conquest in the Twelfth Dynasty, in erecting the earlier of the two temples. It is interesting to observe that their names are nearly all compounded with that of the Theban god Montu, showing that these ancient colonists of Nubia were residents of Thebes.

By Saturday, January 13, we had finished our records of the Halfa temples. They included a complete set of photographs of all the reliefs and inscriptions. In all cases these photographs were carefully collated with the walls, or independent copies of the inscriptions were made in addition to the photographs. Measurements were also taken for a complete ground-plan of the Empire building, in order to locate upon it all the documents reproduced.

The voyage down river was much interrupted by the powerful north wind. It was not until midday of January 15 that we

FIG. 9.—Reliefs of Ramses II Aksheh Temple. The brick arch is Coptic.

sighted the scanty remains of Ramses II's temple at Aksheh, only fifteen miles from Halfa. A dedication of one of the doorposts shows that this temple was sacred to the worship of Ramses II himself. His person, as worshiped here, is called in the dedication "His-Living-Image-on-Earth," a designation of a living, apotheosized king, which is known elsewhere also. Only the inscriptions of the rear wall of the first hall are preserved. The above dedication is on this wall. The south end of the wall bears a list of the southern conquests of Egypt, while the north half is inscribed with a similar list of northern conquests, as we find them a number of times in other temples. The southern list is well preserved, but the northern is badly incrusted with salts, and very illegible. The presence of Naharin, Arvad, and Kadesh, however, shows that it was of the conventional order. Beneath a relief on the north half of this wall is one of those enigmatic, charade-like inscriptions, such as Devéria observed at Leiden many years ago.

We went on to Faras the same day, and the next morning visited four Coptic tombs in the cliffs three-quarters of an hour behind the village. They contain painted Coptic memorial records, which would be worth copying; but our particular mission, and

FIG. 10.—Niches of the Nubian Viceroy Peser at Gebel Addeh.

the serious delays which we had suffered, forced us to devote our time to the older and more valuable temple documents. We returned to the dahabiyeh again just above Dendân (east shore), the last town in the Sudân as you descend the river.

Just after midday of the sixteenth we reached the well-known memorial niches in the Gebel Addeh, in the eastern cliffs, opposite the north end of the island of Shatâwi. The northernmost belongs to the viceroy of Nubia, Peser, heretofore supposed to have been in office only under Harmhab. This niche, however, was made while Peser was serving under Eye—a fact which is of importance as showing that Eye, one of the weaklings at the end of the Eighteenth Dynasty, doubtless immediately preceded

FIG. 11.— Hall of Harmhab's Cliff-Temple at Gebel Addeh.

Harmhab. Another observation is furnished by the southernmost of these niches, which is also a memorial of Peser. Here he bears the title "governor of the gold-country of Amon," showing that already at the close of the Eighteenth Dynasty Amon had gained such power as to possess his own mining region in Nubia. This carries his gold-country 145 years back of its first appearance as hitherto known (*Ancient Records*, III, § 640). Its early appearance here has a twofold interest: first, it shows that the wealth of Amon was founded by the great emperors of the Eighteenth Dynasty, and not by Ramses III; second, it proves that Harmhab restored to Amon the great wealth, of which Ikhnaton's revolution had deprived the god.

On the same day, in the dusk that merges so quickly into starlight on the Nile, we dropped down from the niches of Peser to the cliff-chapel of Harmhab in the north end of the Gebel Addeh. It consists of a columned hall, with a room on either side and a holy place behind it. The entire interior is sculptured with splendid reliefs, like those of the best period of the Eighteenth Dynasty; but these have been hacked out and covered with mud plaster bearing Christian paintings by the Copts. Through dim outlines of St. George or St. Epimachus, riding down the dragon, one discerns the outlines of the four Horuses to whom the temple is dedicated, while a figure of the Christ looks down from the ceiling among much tawdry Byzantine decorative design. The religious inscriptions contain much of geographical importance, but the reliefs, once so beautiful, have suffered sadly. We secured all the inscriptions in the place, except those on the south wall of the hall, which are completely vanished, or lie so deep under impervious plaster as to be totally invisible.

In the afternoon of January 18 we reached the great goal of our winter's work, the vast temple of Abu Simbel. It is not only the largest temple in Nubia, and one of the most remarkable buildings in the world, but is also a storehouse of numerous historical records. It is not necessary here to describe a sanctuary so well known, or to attempt to picture the imposing beauty of its mighty front, along which rise the colossal, seventy-two-foot statues of Ramses II, its great builder. Between the statues the main doorway gives access to a series of spacious halls which penetrate one hundred and eighty feet into the mountain; for the entire structure is hewn from the cliffs of Nubian sandstone, and even the

Sun-Temple   Hathor-Temple

Fig. 12.—General View of the Temples and Monuments of Abu Simbel. Looking westward from a sand-bar in mid-river.

massive colossi of the front are still engaged with the face of the mountain out of which they have been wrought. So long has this temple endured, and in such a fine state of preservation, that the visitor leaves with the impression that it is as enduring as the mountain from which it has emerged. I have everywhere been greeted with incredulity, when I have stated that the temple of Abu Simbel is doomed. The smaller statues grouped between the colossi of the front. are rapidly perishing, having lost noses, whole faces, feet or toes, or other projecting portions, in very recent times. The present very effective administration of antiquities, under the energetic hand of M. Maspero, has long ago forbidden tourist visitors to climb upon the colossi, but the feet of the great statues were badly worn off by such climbing before this prohibition. All are familiar with the imposing fragments of the third colossus, tumbled in gigantic ruin on the visitor's left as he enters the temple. Its neighbor on the right of the door is about to suffer the same fate. A bad fracture appears, running from the bottom at the front diagonally upward and backward through the legs and body. All portions above this crack are resting on an inclined plain of about 45°, and they are slowly shifting downward. Some day, not far distant, they must come crashing down. In the interior the state of affairs is no better. A large piece of the massive architrave on the north of the nave in the main hall, a fragment weighing some tons, fell to the floor only a few weeks before our arrival. The second pillar on the same side in this hall is about to lose a large fragment from its northeast corner, a section anciently inserted by Ramses II's architect. The colossal Osirid statues of this Pharaoh, ranged against these pillars on each side, show many loose fragments, which need but a touch to send them to the floor, while the walls and doorways exhibit numerous fractures, which mean trouble in the near future. Indeed, a huge piece has fallen from the doorway of the second hall. It will be seen, therefore, that this superb temple is going rapidly to destruction, and it is not probable that the disasters which threaten the place can be averted by any work of restoration. Certainly any structure for the support of the fractured colossus would be worse disfigurement than the wreck of the statue itself.

As the interior of this temple is without any light save that which enters at the door, we were here confronted by those

FIG. 13.—Front of the Sun-Temple of Abu Simbel.   From the cliffs on the north of it.   Our dahabiyeh the first on the left.

problems of interior illumination of which we have above spoken. I will confess to some sinking of heart as we went the rounds of the vast interior and counted the scenes and incidents in the Kadesh campaign of Ramses II, which cover the entire north wall of the great first hall. Then, besides this sun-temple, there are also the birth-house immediately on the south, and a few hundred feet northward the temple of Hathor. Scattered

FIG. 14.—Façade of the Sun-Temple of Abu Simbel. Looking westward.

up and down the cliffs in the immediate vicinity of the temples are no less than sixteen large and small historical stelae, only one of which (the Blessing of Ptah) is inside the temples. This one, the Blessing of Ptah, is in the first hall of the largest or sun-temple. A number of these stelae are among the most important historical documents of the Nineteenth Dynasty. Of this great body of documents no uniform publication exists. Many of the more important have been fragmentarily published, but such fragments are scattered through a large number of works, and there is no single volume or series of volumes in which the records of Abu Simbel are collected in complete form for all time, in accordance with modern epigraphic methods.

5.—The Southernmost Colossus at Abu Simbel.   The native sits on the lap of the fallen statue.

The largest of the stelae is that celebrating the marriage of
Ramses II with the daughter of the king of the Hittites. It stands
at the south side of the court before the sun-temple and is some
fifteen feet high. Above is a relief depicting the Hittite king
and his daughter, received by Ramses II, and below is an inscription glorifying the event, containing forty-one lines, each about
eight feet long—a total line some four hundred and twenty-

FIG. 16.—Stelae of Setan at the South End at Abu Simbel.

eight feet in length. The lower three-quarters of this great
document are very badly weathered, and, in the glare of light
which here is reflected from all directions, many places appear
to be only bare and weathered sandstone. Such surfaces demand
an oblique light from one direction at a time and the complete
absence of all light from all other directions. It was necessary,
therefore, to resort to the devices described above (Fig. 3), and as
they have never been applied to this stone before, the result was
a large number of new readings. Of course no account of these
can be given in this connection, but one curious new word is
worth noticing here. Ramses is praying that the winter jour-
ney of his Hittite visitors, as they pass through the northern

countries on their way southward to Egypt, may be free from
"rain and s-r-ḳ." In two different places the two words occur
together, showing that their association is something natural and
common. The new and unknown word is evidently the Arabic
ثلج, Hebrew שֶׁלֶג, meaning "snow." It was curious indeed to
come to snowless Nubia to find such a word for the first time.

FIG. 17.—Photographing Stelae of Setau (see Fig. 16) from Masthead of Dahabiyeh.

Some of the other stelae are now high above the river, and it
is probable that a ledge of rock bearing a highway along the face
of the cliff immediately under these stelae, has now disappeared,
leaving them high above our reach. Several at the south end are
nearly forty feet above the river in the almost perpendicular face
of the cliff. By fastening together long ladders we succeeded in
reaching these for copying, but the offset straight out into space,
necessary for photographing these, could only be secured by
climbing to the masthead of a dahabiyeh. Others required the
building of tall scaffolding, with sufficient projection to secure the
necessary offset.

FIG. 18.—Photographing Stela of Nubian Viceroy Ini at Abu Simbel.

FIG. 19.—Stela of Nubian Viceroy Ini at Abu Simbel, as photographed from high scaffolding (Fig. 18). The foreshortening is due to the slant-ing face of the cliff.

The interior of the great sun-temple was a slow and arduous task. Day after day the tall scaffolding rose and fell, as we passed slowly across the walls, the camera recording for us a thousand data, which would have required weary weeks for the draughts-man to put upon paper. Undoubtedly the modern draughtsman is producing very full and accurate facsimiles of such pictorial records as those of the battle of Kadesh in the great hall at Abu

FIG. 20.—Excavating a Sand-covered Stela of Ramses II at Abu Simbel.

Simbel, but he has not visited this hall. The fantastic ideas generally current regarding the appearance of the Sherden—probably Sardinian mercenaries serving in Egypt in this age—owe their origin to the published drawings of Ramses II's Sherden bodyguard made by Rosellini's draughtsmen. I wish distinctly to disclaim any reflection upon the self-sacrificing devotion with which such records were made by our predecessors in the early decades of Egyptology. It is this early work which has slowly enabled us to go farther, and it is our good fortune to enjoy the use of modern mechanical methods of accurate and rapid reproduction not possessed by the great pioneers of our science. With unfailing enthusiasm they applied to the reproduction of such

1.—Corner of the Great Hall in the Sun-Temple of Abu Simbel. Holy of Holies visible at the left.

monuments all the means then at their disposal; but it is now our
duty to do the same with the vastly improved processes now at
our command.   In placing side by side the work of the old
draughtsman, and that of the modern camera, I only desire to
show that the work of producing the final record of such monuments still lies before us, and that their rapid dissolution demands

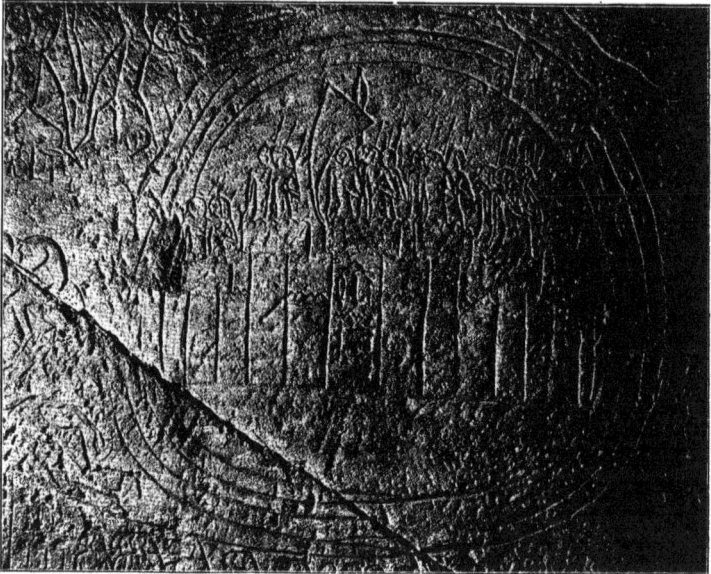

FIG. 22.—The Fortified City of Kadesh on the Orontes which flows around the Walls;
Hittite Troops with Banner in the Turrets.   From the reliefs of the Battle of Kadesh at
Abu Simbel.

immediate application to the problem, if we are to possess
anything more than the insufficient records of primitive
Egyptology.

It was with great enthusiasm that we undertook the application of these methods to the wall bearing the records of the first
battle in history of which we can trace the strategic disposition
of the opposing armies, and discern the fact that clever and mis-
leading manipulation of forces masked behind hills and city walls
was an art already long practiced and highly developed.   Here
we see Ramses separated from the great body of his army by a
deft flank movement to which he had exposed himself.   All is

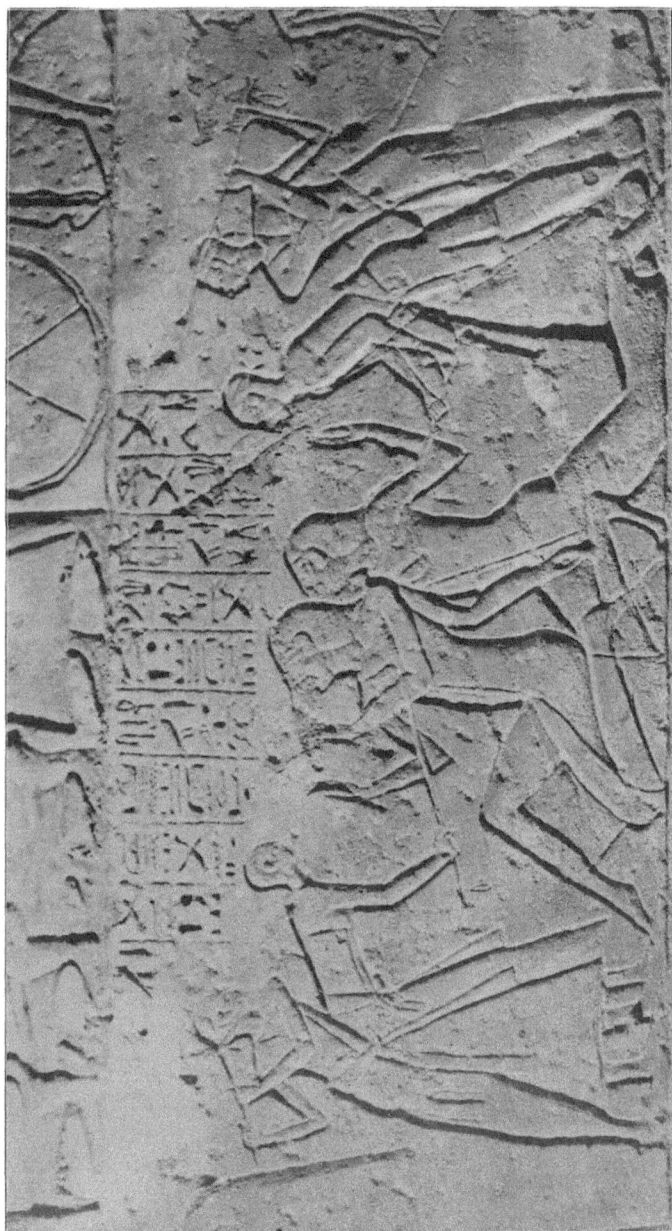

FIG. 23.—The Beating of the Hittite Spies. The inscription above them reads: "The arrival of the scout of Pharaoh, bringing the two scouts of the vanquished chief of Kheta (Hittite country) into the presence of Pharaoh. They are beating them to make them tell where the wretched chief of Kheta is." From the reliefs of the Battle of Kadesh at Abu Simbel.

depicted upon the wall: the council of war, in which tortured spies disclose the whereabouts of the enemy and Ramses learns of his fatal isolation; the camp surrounding the royal tent where the council is held; the king's tame lion bound behind his master's pavilion; the onset of the enemy as they pour into the camp; Ramses' desperate offensive as he endeavors to save the day; the messengers sent to bring up the distant divisions; their forced march to save the king; and the final triumphant inspection of prisoners and slain when the ordeal is safely past—all these have found place upon the wall. It is in these monuments recording Egypt's collisions with the northern world that Europe first emerges in written documents, and the importance of exhaustively accurate reproductions of physiognomy, costumes, and arms will be evident.

Such records so elaborately executed, the size of the temple, the other temples beside it, the great historical stelae, and the numerous monuments of officials on the neighboring cliffs, would indicate that Abu Simbel was a place of great importance in the Nineteenth Dynasty. The large or sun-temple is dedicated to the three great gods of the Empire—Amon, Re, and Ptah—and to the king, Ramses himself, who is even depicted a number of times in the reliefs as king offering sacrifice to himself as god! It was thus an important state temple; but the remnants of the town which must have existed in the vicinity have totally disappeared, and its location is now a matter of pure conjecture. The sands invading the valley from the west have completely covered any traces of it which might have survived. Indeed, even the temples themselves are protected with difficulty from the yellow drifts which engulf everything in their course. The front of the great temple has been cleared a number of times. In 1892 Captain Johnstone, R. E., built two walls designed to prevent further encroachment, but these hopes proved delusive, and the northern half of the front is again encumbered, while the advance of the tide is steady and irresistible. It has now reached the door again. (See Figs. 12–14.)

The small temple immediately south of the sun-temple has, since Miss Edwards' time, been called the temple library. For this there is no evidence; on the contrary, the inscriptions in the place distinctly call it a "birth-house." It is, therefore, the oldest example of the chapel of the divine mother-consort and her son, of which we have a number of examples in Ptolemaic times. The Hathor temple on the north was erected to this goddess by Ramses II and his queen Nofretiri; it is not sacred to the queen, as is often stated. (See Figs. 12 and 24.) Like those of the birth-house, its inscriptions are exclusively of religious interest. We took complete records of both these temples.

For many years there has been a report current among the natives about here that there is an unknown temple far out in the desert behind Abu Simbel. Various explorers have examined the neighboring desert in the hope of finding it, but thus far without success. A native told me that Maspero took eight camels many years ago and scoured the desert for three days in a fruitless search for this fabled temple. I have never asked M. Maspero regarding the truth of this story. However, a villager approached me, in response to inquiries which I sent among them, and assured me that he had found this temple, since M. Maspero's search, and that he could lead me thither. Securing the necessary camels, therefore, I struck out into the desert with this man one morning, bent on seeing the phantom temple. In such a quest as this, it must not be forgotten that the natives of Nubia apply the term *birbeh*, "temple," to any structure whatever. They even call a niche in the cliffs or an ordinary cliff-tomb a *birbeh*. Entering a wady a quarter of a mile south of the sun-temple, we left the Nile, and proceeded northwestward for some forty minutes, when we had ascended to the desert plateau. At this point we turned directly northward. We journeyed thus for two hours into the desert, diverging from the Nile at an angle of 45°. My guide then pointed to what I must confess looked much like a distant building rising out of the sand in the north. Eagerly I pushed on for closer inspection, full of curiosity as to this mysterious sanctuary of the desert. As we drew near, the supposed building resolved itself into an isolated crag of rock, projecting from the sand, and pierced by two openings which passed completely through it, so that the desert hills on the far horizon were clearly visible through them. One of these openings very much resembles a door, and, to complete the delusion, it bears on one side a number of prehistoric drawings—two boats, two giraffes, two ostriches, and a number of smaller animals—which might be easily mistaken by a native for hieroglyphic writing. There can be no doubt that this curious natural formation and the archaic drawings upon it are the source of the fabled temple in the desert behind Abu Simbel. On our way home we went east toward the Nile, and passed through a camp of the dervishes, who had perished there on their invasion of the region by Wad en-Nugûmi, before their defeat at Toshkeh by Greenfell in 1889. The plain is white with their bones, and the ground is strewn with utensils and camp rubbish, while here and there, far out in the sands, lies

FIG. 24.—Hall of the Hathor-Temple at Abu Simbel. See also Fig. 12.

a grinning corpse still wrapped in ragged garments, fluttering in the wind, or half covered by the drifting sands of the waste.

A few days after this excursion we had nearly reached the close of our work at Abu Simbel. The Hathor-temple gave us some trouble. The halls are much smaller and lower than those of the sun-temple and in illuminating for time exposures the smoke drove along the ceiling before the exposure was finished, and descending the side walls appeared in white clouds within range of the instrument, producing a bad overexposure wherever it appeared. We found it very difficult to exclude this smoke from the field of the instrument, but finally succeeded in creating sufficient draught to carry the worst of it away. One of the stelae

FIG. 25.—Looking Up-River from the Byzantine Fortress of Kasr Ibrim. The walls are visible on the height at the left; our dahabiyeh at foot of the cliffs. A sand-bar in mid-river makes it appear narrow. The opposite shore-line is marked by the fringe of trees.

just north of this temple also made some trouble. It required
our highest scaffolding, which was none too stable. Just as all
was in readiness and Koch was exposing, a sudden whirlwind
swung round the cliff with terrific force. It tore the dahabiyeh
from her moorings, and only the timely warning of a sailor saved
Koch and our large camera from being carried off the high tressel
to the rocks below.

On February 24 the work at the great Abu Simbel group was
completed. None of our party will ever lose the impressions
gained during the weeks spent under the shadow of the marvelous sun-temple. In storm and sunshine, by moonlight and in
golden dawn, in twilight and in midnight darkness, the vast
colossi of Ramses had looked out across the river, with the same
impassive gaze and the same inscrutable smile. I have turned
in my couch in the small hours and discerned the gigantic forms
in the starlight, and been filled with gladness to have had the
opportunity of doing anything to preserve the surviving records
of the age that produced them. What are the privations of travel,
and prolonged separation from home and friends, if only something can be accomplished to render permanent in adequate records which shall preserve them for another thousand years, such
matchless works!

Delayed by north winds, it was not until the morning of February 26 that we reached the grottoes of Kasr Ibrim. There are
five of these cut into the face of the cliffs on the east side of the
river and four of them are inscribed. It was evidently a custom
of the viceroys of Nubia to record the successful collection of the
Nubian tribute at this place, as a memorial of their administration of the country. Scenes of offering to the gods of the south
gave to such memorials also religious significance. It was here
that we found a new record of the Nubian tribute of the Eighteenth Dynasty. It records how the king sat in state on the
great throne (_tnt'yt_) in Thebes, while the tribute was brought
before him by long lines of carriers. It gives us an impressive
picture of this annual scene in the great capital of the Empire,
besides furnishing some economic data of value. The amounts
of the various articles of the tribute are given in man-loads, which
is an entirely new method, not elsewhere found on the monu-
ments· Thus we find here·a thousand men bearing ebony, ten
men leading live panthers, and some two hundred and fifty laden

with perfumes and aromatic woods The different articles of the
tribute are: gold, minerals (ḥm'gt), ivory, ebony, perfumes aromatic woods, panthers, hounds, oxen, and cattle One item is
illegible at the head of the list They are brought forward by
2,667 men, the total being given by the scribe at the end of the
document, while the number bearing a given commodity is each
time added after the name of the article.

FIG. 26.—Reliefs from the Tomb of Penno at Anibeh. The upper row shows the Viceroy of Nubia bowing in audience before the Pharaoh (at the left). In the same row (right
end) stands the statue given by Penno, while before it with staff in hand is the Viceroy,
followed by one of his officials.

By eleven in the morning of February 27 the grottoes of Kasr
Ibrim were finished, but we were unable to make any progress
northward against a tremendous north wind, so that we went
across country on the west shore to the tomb of Penno behind
Anibeh. It contains an interesting document recording the dona-
tion of five pieces of land, the income of which is to be used for
the maintenance of a statue of Ramses VI in the neighboring

temple of Derr. Both statue and lands are the gifts of the loyal Penno, and the royal recognition of his loyalty is recorded also in the tomb. The King sent no less a person than the viceroy of Nubia himself, with two silver vessels which the viceroy carried to Ibrim and presented to Penno. All these events and the deed of gift for the land are duly recorded or depicted in the tomb. The tomb chamber containing the records is hewn into the cliff, but artificial illumination made it possible to complete good negatives of the entire interior by the time darkness had set in, on the evening of the twenty-seventh.

The next day was spent in recording a victorious stela of Seti I, in which he is glorified in such conventional language that it is impossible to discern what particular victory of his over the Nubians may be meant. The stela is engraved in the eastern cliffs in the first promontory north of Kasr Ibrim. Early the next morning we reached the village of Ellesiyeh, behind which is the cliff-chapel of Thutmose III. It must have been a beautiful monument when in a good state of preservation, but its one room has been ruined by bats, and has suffered too from being used as the back room of a dwelling, the roof of the front portion of which was supported on timbers inserted into a row of holes still visible along the top of the façade. I copied the religious inscriptions of the interior, as it was impossible to photograph them. The stela of Thutmose III's fifty-first year outside, on the right of the door, contains, as far as legible, no more than conventional epithets in praise of the king. Its pendant on the other side of the door is of similar content. Its date is uncertain, and may be either year 26, or 42, or possibly 51. The front of the chapel bears a number of interesting graffiti of officials who have visited the place on some errand for the government or the Pharaoh, and have improved the opportunity to record their names and titles on the rock.

The next morning found us at the cliff-temple of Ramses II at Derr. There is no more graphic evidence of the decline of the provincial arts under Ramses II than this temple of his at Derr. It is easily conceivable that with such a large number of vast buildings in process of construction, it was impossible for the royal architects to find master-workmen to put in charge of such structures in the province. But, giving this consideration full weight, it still remains a very significant fact that such large

FIG. 27.—Hall of the Temple of Derr, showing Bend in Architrave.

FIG. 28.—Osirid Pillars in the Hall of
the Temple of Gerf Hus ê.

and important temples as those of Ramses II at Derr and Gerf
Ḥusên should display such extremely slovenly work, both in
architecture and in sculpture. So badly was the work directed here
at Derr, that the axis of the main hall, as the excavation into the
cliff proceeded, had to be changed, and there is therefore a sharp
turn in the architrave, which can be clearly seen in the illustration (Fig. 27). The pre-eminence of Abu Simbel among the
Nubian temples is not alone due to its vast dimensions; the
greatest sculptors of the age must have worked there, while at
Derr the men who wrought the royal figures were little better
than ordinary stonecutters. The historical records here are all
in the first hall, which was only cut partially from the cliff. The
inner row of pillars still bears architraves which are hewn from
the cliff in place, but architraves of the other, *outer* pillars were
detached blocks, as were also the upper portions of these pillars.
The upper and outer part of the hall was therefore masonry
blocks. These have now disappeared, and with them a large
portion of the historical records on the walls of the hall. What
remains is so frightfully weathered that but little can be discerned.
It is evident, however, that the reliefs depicted the conventional
achievements of the Pharaohs. This temple never contained
any records of the Kadesh campaign;[1] the condition of the building, when Champollion visited it, was exactly the same as at the
present day, and no loss of historical records since Champollion's
visit has taken place. The documents which we copied and photographed are exactly those described and located by Champollion,
whose account of them, and marvelous insight into their content
and meaning, must always rouse the deepest admiration.

In the afternoon of March 3 we had reached the temple of
Amada (Figs. 35 and 39). This building is the perfection of all
that the temple of Derr is not. The exquisite reliefs are wrought
with a delicacy and taste, and colored with an elaborateness and
precision, found only in the best work of the classic Eighteenth
Dynasty. It is remarkable that this beautiful sanctuary has found
so little appreciation, for there is nothing at Thebes any better.
I do not refer to the architecture so much as to the superb reliefs
with which the interior of the building is throughout adorned.
They were plastered over by the Copts, who used the place as a

---

[1] Wiedmann (*Gesch.*, 434) bases the existence of such Kadesh records here, on Champollion, who, however, makes no statement that he saw them here (*Notices descr.*, I, 86–95).
My *Battles of Kadesh* (P. 8) is to be corrected accordingly.

church, and this has resulted in the preservation of the colors. These are now being rapidly destroyed by the neighboring natives, who are cutting out the cartouches, the heads, and the finest of the hieroglyphs for sale to the tourists, who of course readily buy. It is much to be hoped that the government may be able to appoint a watchman in this temple. A native watchman costs but a pound a month, and this insignificant expense will save for us one of the

FIG. 29.—The Temple of Amada. Rubbish from our excavations in the foreground.

greatest monuments of ancient Egypt. While we were at work here, the mamûr of the district came over from Derr to see what we were doing. I told him of the ragged holes cut in the beautiful walls for the sake of taking out pieces to sell. The mamûr thereupon offered the ingenious suggestion that these holes had been made by wild beasts in the effort to excavate dens for themselves. When I took him into the temple and showed him the holes, however, he was quite willing to abandon his theory and accept the explanation that they had been made by the natives of the vicinity. His visit frightened away half our workmen, and it is to be devoutly hoped that it will also withhold the native vandals from further destruction of the temple.

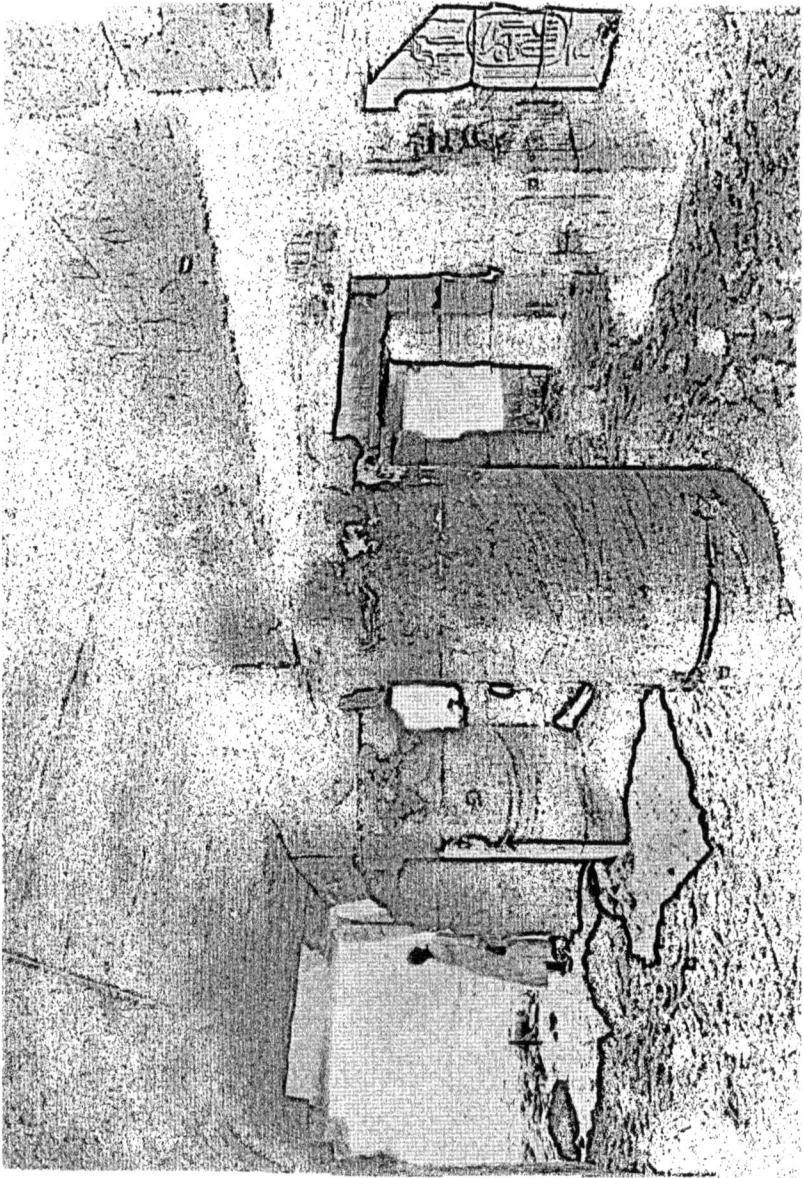

After a few days' work at Amada it became evident that we could not make a final edition of the inscriptions without removing the sand from the first hall. We put two of our sailors at work sinking a shaft through the sand alongside one of the pillars, in order to penetrate to the pavement and to ascertain the depth of the stuff to be removed. They reached the floor a little over two meters down. The excavation disclosed the fact that

FIG. 31.—Leveling Rubbish from the First Hall at Amada.

beneath the sand there was a stratum of rubbish about sixty centimeters deep. It was evident that this rubbish had never been cleared away in modern times. We secured a gang of villagers from the other side of the river, through the efforts of the village sheik, and in a little less than a week we had cleared the first hall and removed enough sand from the second hall to expose the walls to the lower edge of the reliefs. In all the other rooms our sailors removed the sand along the walls and piled it in the middle of each room. All the records in the temple were thus laid bare. We found the Coptic plaster still undisturbed for a meter from the floor along all the walls and on all the piers of the first hall. As this plaster was not painted or inscribed, we cleared it off, and this disclosed six new inscriptions. Two were records of the viceroy of Nubia, Heknakht; two were memorials of the mercenary commander of Nubian archers, Epyoy (Py'y) recorded by him in honor of the queen Tewosret and the chief treasurer Bay. The fifth recorded the second jubilee of Thutmose IV, and the sixth was a Coptic inscription.

We also came upon a number of fragments of inscribed private monuments in the rubbish of the first hall. The most important of these was a pyramidion of the Nubian viceroy Mesuy, nearly a meter high, but with the apex gone. It had been mounted upside down by the Copts as a seat beside the door. Mesuy's name has been expunged from all five records, which he has left in this temple. He must therefore have suffered political overthrow. It is he who recorded the long thirteen-line inscription of Merneptah in the doorway of the first hall, and it must have been the political turmoil at the death of Merneptah, in which Mesuy's party fell, and he lost his office and power. It may, indeed, have been the above Epyoy who expunged the unfortunate Mesuy's name. The long inscription of Merneptah is known only in a casual copy by Bouriant, in which even so important a matter as the date was not discerned. Now, this date is the same as that of the beginning of Merneptah's Libyan war, so that the first part of the inscription refers to that war, to which it contributes some interesting new facts. It is this document which calls Merneptah "Binder of Gezer," showing that he had campaigned in Palestine, as claimed by his great stela, which represents him as having desolated Israel (*Ancient Records*, III, §§ 602 ff.).

The fine stela of Amenhotep II, engraved on the rear wall of the Holy of Holies, we of course recorded with special care. The walls of some of the rooms showed holes, into which lamps or lamp-stands of some kind had been inserted. Melted fat or resin had run down the wall beneath such holes, and slowly formed a dark, hard, gummy, or resinous enamel, which filled up hieroglyphs and made them totally illegible. It was exceedingly difficult to remove this tough, elastic surface, which was like the cartilaginous rubber surface of insulated wire. The lower portion of the great Amada stela has suffered much from such disfigurement. On removing it, however, a number of new readings came out.

We found that the removal of the Coptic plaster in the rear rooms exposed some very fine painting—a fact which made us

FIG. 32.—First Hall of the Temple of Amada after Excavation. Viewed from *r* (on plan), including fro
llar 4 (at right) to 1 (at left) and 12 (at right) to 15 (at left). The depth of the excavation may be seen by co
ring the cartouches or royal ovals on the square pillars in Fig. 30 with those on the same pillars above (Fig. 32

greatly regret that we were not equipped to do water-color work. It is devoutly to be hoped that someone capable of copying such temple paintings in color may visit the place and preserve in this way some of the best of these colored reliefs. Another disadvantage was the fact that orthochromatic photographic plates do not keep in a warm climate like that of Nubia. Every photographic expert whom I consulted strongly advised against attempting to use orthochromatic. plates on such an expedition. We therefore had no plates highly sensitive to color with us. We found, also, that the magnesium flame, which serves admirably for the illumination of reliefs, is not so effective in bringing out on the plate such details as are done on the wall solely in color. Much could therefore still be accomplished to make a more adequate record of the superb painted reliefs in the Amada temple.

The history of this building is of value for the history of the dynasty which produced it. Amenhotep II's great stela in the Holy of Holies furnishes a part of that story of the building which may serve as a basis upon which to reconstruct the whole. It reads as follows:

Behold his majesty [Amenhotep II] beautified [that is "decorated"] the temple which his father, King of Upper and Lower Egypt, Menkheperre [Thutmose III], had made for his fathers, all the gods, built of stone as an everlasting work. The walls around it are of brick, the doors of (cedar of the best) of the terraces; the doorways are of sandstone, in order that the great name of his father, the Son of Re, Thutmose [III], may remain in this temple for ever and ever.

The majesty of this Good God, King of Upper and.Lower Egypt, Lord of the Two Lands, Okheprure [Amenhotep II], extended the line and loosened the —(?) for all the fathers (the gods), making for it a great pylon of sandstone, opposite the hall of the festival-chamber (?) in the august dwelling, surrounded by columns of sandstone as an everlasting work. (*Ancient Records*, II, §§ 794, 795.)

The nucleus of the temple (see plan, Fig. 33, halls *BCDEFG*) was erected by Thutmose III, evidently late in life, perhaps during his short coregency with Amenhotep II. When he died, as the above excerpt shows, he left the building without wall decorations. These were supplied by Amenhotep II, who devoted half of each chamber to relief scenes of a religious nature, in which his father figured, while the other half was similarly devoted to himself. This also accounts for the fact that Amen-

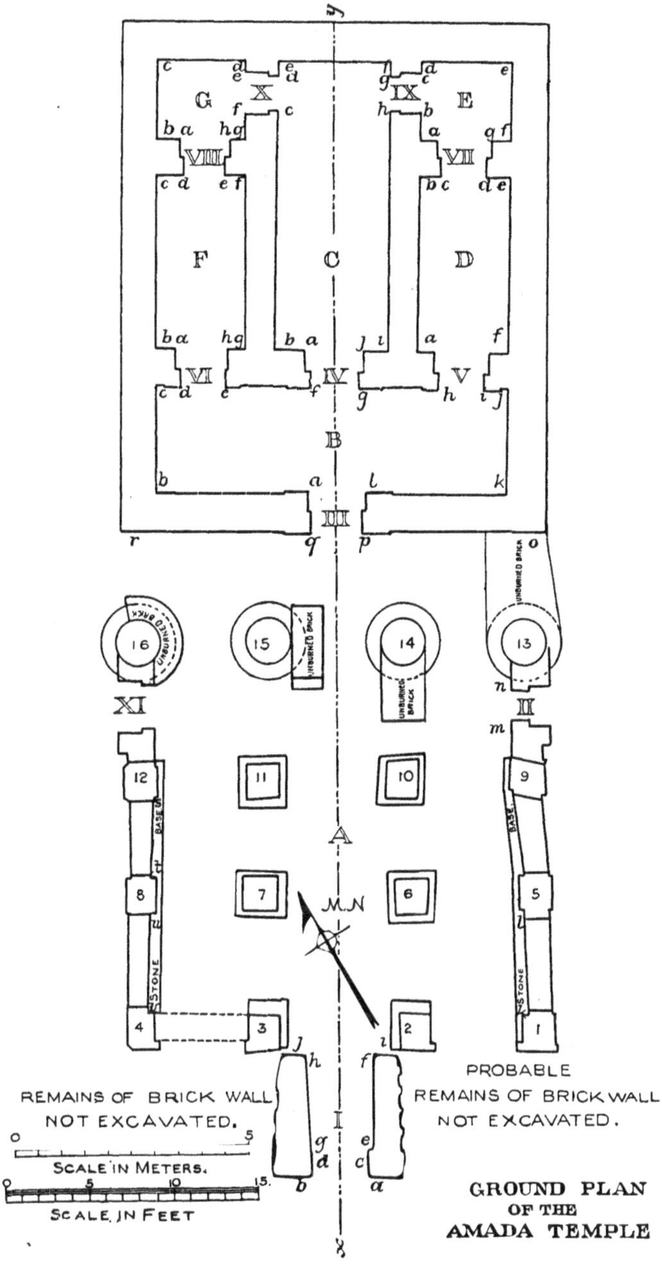
GROUND PLAN
OF THE
AMADA TEMPLE
REMAINS OF BRICK WALL
NOT EXCAVATED.
PROBABLE
REMAINS OF BRICK WALL
NOT EXCAVATED.
SCALE IN METERS.
SCALE IN FEET
UNBURNED BRICK
BASE
STONE
M.N
A
B
C
D
E
F
G

FIG 33

hotep found the rear wall of the Holy of Holies (room *C;* wall *e-f*) unoccupied, so that, in his third year, he could engrave upon it his great record of his conquests in Asia and Nubia, from which we have quoted the above remarks regarding the building.

Before this building (*BCDEFG*) was a colonnade of four columns (13–16), likewise erected by Thutmose III; for columns 13 and 14 bear his name, and column 14 has a dedication reading: "Thutmose III; he made it as his monument for his father Re-Harakhte, making for him an august colonnade." Columns 15 and 16 were inscribed by Amenhotep II, who put a similar dedication on column 15. He therefore shared these columns with his father, as he did the walls of the chambers and halls within. But the architrave running transversely along these columns, from 13 to 16, bears the name of Amenhotep reaching from one end to the other, or, rather, beginning in the middle and extending in duplicate each way to the two ends (at 13 and 16). None of the pillars (1–12) was standing at this time. Before the colonnade (13–16) was a court, to which Thutmose III had perhaps built a sandstone portal. Amenhotep, in the great stela (see above) claims that he built this portal, a *bḫn·t*, possibly "pylon," as he calls it. If so, he was magnanimous enough to share the inscriptions upon it with his father, for the eastern half is inscribed with the name of Thutmose III, like the eastern columns (13, 14). This portal was, of course, the gateway of the brick inclosure-wall which surrounded the building. This brick wall, as Amenhotep above admits, was constructed by his father. Such was the temple at the death of Amenhotep.

Thutmose IV, who succeeded him, then extended the porticus (13–16) into a pillared hall, erecting 12 pillars (1–12) supporting four architraves (1–9, 2–10, 3–11, 4–12), at right angles with those of the porticus (13–16). These latter architrave blocks (13–16) had to be cut out for the insertion of the new ones at right angles. Before this was done, however, it was necessary to suppress the name of Amenhotep II extending along the front of these blocks (13–16). This was done by turning the blocks with the inscribed face inward—that is, toward the rear. Sufficient respect was shown Amenhotep II, however, not to turn the blocks over, which would have put his name upside down; but the entire length of the architrave was turned end for end, so that the end formerly on column 16 now rested on column 13.

Thus the name of Amenhotep II is still to be read, right side up, on the back of the architrave (13-16), while upon the former back thereof, now the front, is the name of Thutmose IV.

The building was thus complete. The purpose of this last hall (*A*), erected by Thutmose IV, is of great importance for his reign. It was erected in celebration of his second jubilee, or, as the record oddly puts it, "the first occurrence of repeating the

FIG. 34.—Record of the Celebration of the Second Royal Jubilee of Thutmose IV. Found in the nave of the first hall (pillars 2, 6, 10 and 3, 7, 11) at Amada.

jubilee" (*sp tpy whm ḥb śd*). Our excavations disclosed beneath the reliefs on the pillars facing the nave (2, 6, 10, 3, 7, 11), six times repeated, the inscription shown in Fig. 34. The purpose of the hall is thus clearly indicated. As the first jubilee was celebrated thirty years after the king's appointment as crown prince, and the second three years later, it becomes evident that Thutmose IV reigned until thirty-three years after his father had appointed him as crown prince. It is probable from the paucity of his monuments that Thutmose IV did not enjoy a long reign. Hence he must have received his appointment from his father, far back in the reign of the latter. This fact throws an interesting light on the story engraved on the great Sphinx Stela at Gizeh, in

FIG. 35.—The Great Bend of the Nile at Korosko. Looking up-river. Amada is on the farther shore where the river disappears in the distance.

which he is promised the kingship by the god. If this tale is at all historical, the incident which it narrates (*Ancient Records*, II, § 810 ff.) must have taken place very early in the youth's life, and his appointment as heir to the succession must have occurred soon after.

It is greatly to be hoped, as we have said, that the government may station a watchman at this temple immediately. Repairs are also urgently needed here. In hall *A* it is only a matter of a short time when at least four of the architrave blocks must fall, bringing down with them a large part of the roof. At present the roof lacks but eight blocks; two over the nave, two at the northeast corner, two at the southeast corner, one over door *I* (now resting on the top of that doorway), and one at the southeast corner. The two over the nave were undoubtedly broken out by the Copts, who erected a pitiful brick dome over the opening thus created. If the sand were cleared entirely from the inner chambers, and wooden covers placed over the windows in the roof, the interior could be kept clear of sand, the roof-windows only being opened for the sake of light when needed by visitors, and closed again by the watchman.

Reaching Korosko on March 18, I spent half a day looking for the inscription of Amenemhet I, recording the first campaign of the Middle Kingdom in Nubia. It was first seen by Dr. Luetge in 1875, and by him was shown to Brugsch; but I was unable to find a native who knew anything about it, and our own search failed to discover it. As our time was now fast slipping away and the warm weather was hard approaching, we gave up the search, in view of the fact that the inscription consists of but a few words.

The morning of March 20 disclosed the temple of Sebû'a behind the palms on the west shore. It is so enveloped in sand that but little more than the pylon stands free. Fortunately it contains no historical inscriptions, and practically no documents of any historical importance. I say fortunately, for it is so deeply incumbered with sand that its clearance would be a long and expensive undertaking. The chambers and halls are all full to the ceiling, and the forecourt likewise from the top of the door of the first hall to the base of the pylon. As there is no danger of loss to these sand-covered walls, the few religious inscriptions and scenes which they bear can be safely left until the government shall have cleared up the place and made them accessible This temple of Sebû'a is exclusively the work of Ramses II, and the records within it are but a repetition of those which he has so often placed in his Nubian temples, like the list of his sons and daughters which is found on the back of the pylon facing the court. The names are those well known from the other lists. This is the only pre-Ptolemaic temple, between the first two cataracts, of which we did not secure a complete record.

Fig. 36.—The Sand-covered Temple of Sebû'a.

Having spent the night at the village of Melahad, the morning of March 21 found us passing the villages of Merga, Shema, and El Egêba. Beyond the last we approached a point of rocks which rise at the water's edge on the west shore. With the glass a group of reliefs could be discerned, depicting elephants and giraffes. As the former have been extinct in this region probably as far back as the Old Kingdom, these reliefs very likely antedate that age. Nearly a kilometer farther north, behind the village of El Madîk, there is a second inscribed rock bearing graffiti of Middle Kingdom officials. Three of these are dated, but unfortunately do not add the name of the king to whom the year

Village of El Madîk, Looking Down-River. A typical Nubian landscape: on the left the cliffs along the margin of the d
ouses of the village; the latter separated from the river by a narrow stretch of grain-fields and a fringe of palms; our boa
s.

refers. These inscriptions are on the rocks directly over the
houses of the village (see Fig. 38). Yet when I asked one of
the natives if there were inscribed rocks near, he said "no," and
all the other villagers confirmed his statement. It thus became
evident that the only way to secure these scattered records of
Egyptian rule here is to go out among these burning cliffs and
search for them with a glass.

FIG. 38.—Looking Up-River from El Madîk. Taken from the same point of view as
Fig. 37, but looking in opposite direction.

Unfortunately, however, the work of recording the temple
documents, which was our particular object in this region, had to
be preferred at all costs. There was not sufficient further time
remaining at our disposal, therefore, to conduct an exhaustive
search along both shores of the river. The eastern shore along
this stretch of river undoubtedly would yield some few such
records; we could only employ our progress to the next temple
searching the western cliffs with a glass as the boat passed. Late
in the afternoon of the same day, about a kilometer south of the
village of Molokab, we descried from the deck of the boat, high
up on the western cliffs, a royal inscription (Fig. 39). On land-

ing the rock was found to be covered with graffiti of Middle Kingdom officials, dominating which was a large royal titulary which
must belong to the Middle Kingdom; but I have not yet succeeded
in placing it finally. It evidently is a new king. The little inscription under the cartouch, "The Prophet Khnumhotep," is of the
same workmanship and style as the royal names, and must have
been put there at the same time. Above these inscriptions, on
the rocky plateau, are remains of numerous rough, unhewn stone

FIG. 39.—Name of a New King, Probably of the Age of the Eleventh
Dynasty (2160-2000 B. C.). Found on the western cliffs at Molokab.

buildings. Just north there is a bay in the shore, and above this
bay, on its north side, are remains of a Coptic church with poly-
chrome decoration, and numerous Coptic texts painted in red on
a white plaster ground. Only a few feet of sand-covered wall now
survive, but someone with more time for such incidental work
should attempt the rescue of the Coptic texts.

A few more graffiti were secured during the voyage of the next
day, and we spent the night just south of the Byzantine fortress
of Ikhmindi (Mehendi), which is a little over a mile south of the
temple of Maharraka or Offedineh. This point, anciently called
Hierasycaminos, was the southern limit of Ptolemaic administration. Here, therefore, begin the series of late temples of Graeco-
Roman age, extending to the first cataract. These, however,
did not fall within our programme. We undertook, as I have

FIG. 40.—Graeco-Roman Temple at Maharraka.

FIG. 41.—Looking Up-River from the Summit of the Pylon of the Dakkeh Temple.

said before, only the pre-Ptolemaic—or, as they happen to be, the Empire—temples of this region. While the Graeco-Roman buildings of northern Nubia are of great architectural interest, and it is to be hoped that an exhaustive archaeological survey will be made of them, before they are endangered by the raising of the dam at the first cataract, nevertheless there could be no question in our minds which monuments were the more impor-

FIG. 42.—Looking Riverward from the First Hall of Ramses II's Temple of Gerf Ḥusên.

tant, or more urgently demanded immediate attention. We did no more at Maharraka and the other Graeco-Roman temples, therefore, than to take a few general views and photograph the more important Greek and Demotic inscriptions.

On March 24 we spent half a day at the Graeco-Roman temple of Dakkeh. By midday of Sunday, March 25, we drew in at the temple of Gerf Ḥusên, erected by Ramses II (Figs. 28 and 42). In character it much resembles his sanctuary at Derr, and is not superior to that slovenly structure in architecture or sculptures. The historical reliefs which occupied the side walls of the first court have totally disappeared, as at Derr, and the remaining records in the halls cut from the cliff are exclusively religious in content. The place occupied us but a few days, and by March 28 we were at Dendûr, which we left the next morning·

We were now but forty-four miles from the first cataract, and a good south wind enabled us to run to Shellâl, at the head of the cataract, to secure some fresh provisions, which, owing to the high water caused by the big dam, cannot be secured for many miles above the cataract. Having laid in the supplies on the thir-

FIG. 43.—Temple of Dendûr.

tieth, a strong north wind carried us back to our next temple of Bêt el-Wâli at Kalabsheh, which we reached by mid-afternoon. It is the work of the inevitable Ramses II, but is vastly superior in its sculptures, even to the reliefs of Abu Simbel. In the fore-court on the side-walls are war-reliefs, and records of tribute of great importance, and so situated that they can be beautifully photographed; though it was necessary to do half of them at night by artificial light.

When we had finished this place on April 4, the north wind had risen to a gale which made any progress northward hopeless. We waited a day or two for it to abate, while busily arranging and indexing our records; but as it still continued, we

FIG. 44.—Looking Down-River from the Door of the Temple of Bêt el-Wâli. The wady at the left is filled by the backwater from the Assuan dam forty miles away.

FIG. 45.—The Cliffs and the Temple of Bêt el-Wâli.

cast off and ran south to the Middle Kingdom fortress of Kubbân, of which we made a summary survey. No more could be done without considerable excavation. What we did manage to accomplish was done with the greatest difficulty, for the wind had now strengthened to the heaviest gale we had yet met on the Nile. A camera would not stand for a moment, no matter how well anchored, and surveying instruments could not be kept upright.

FIG. 46.—Tribute of Nubia Brought before Pharaoh. Reliefs from the Temple of Bêt el-Wali.

A trip northward along the eastern cliffs resulted in the finding of a few graffiti of the Middle Kingdom. Across the Nile vast clouds of sand were driving along the horizon, the palms, dimly visible through the murky atmosphere, bowed and heaved, and tossed wild tops, while the yellow river was a mass of white-capped waves. Such sandstorms as these make photographic work on the upper Nile exceedingly difficult, as it is impossible to protect instruments or plates from the penetrating dust, which soon spoils every tight joint, where two surfaces move one on the other. Shortly after this, Mr. Persons, whose health had been

FIG. 47.—Hall of the Bêt el-Wâli Temple.

FIG. 48.—Looking Up-River Across the Graeco-Roman Temple of Kalabsheh.

bad throughout the trip, was completely incapacitated. He was soon in need of medical care, so that we hurried to Assuan, where on physician's advice he was sent to Cairo, and sailed immediately for America. Fortunately our programme for the winter was completed with the survey of Bêt el-Wâli, and we could now make our way to Assuan as soon as circumstances would permit. It was not until April 22 that we had completed the filing of materials and the packing of the outfit to be left in storage at Assuan for use next season. On this date, therefore, our first campaign was concluded.

The following temples, chapels, stelae, and other important monuments were recorded for publication, besides graffiti and less important records:

1. Halfa, Eighteenth Dynasty Temple of Thutmose III.
2. Halfa, Twelfth Dynasty Temple.
3. Aksheh, Nineteenth Dynasty Temple of Ramses II.
4. Gebel Addeh, Nineteenth Dynasty Temple of Harmhab.
5. Abu Simbel, Nineteenth Dynasty Temple of Ramses II.
6. Abu Simbel, Nineteenth Dynasty Birth-House of Ramses II.
7. Abu Simbel, Nineteenth Dynasty Hathor-Temple of Ramses II.
8. Abu Simbel, Sixteen Historical Stelae.
9. Kasr Ibrim, Eighteenth Dynasty Chapel, Thutmose III.
10. Kasr Ibrim, Eighteenth Dynasty Chapel, Thutmose III and Hatshepsut.
11. Kasr Ibrim, Eighteenth Dynasty Chapel, Amenhotep II.
12. Kasr Ibrim, Nineteenth Dynasty Chapel, Ramses II.
13. Anibeh, Twentieth Dynasty Tomb of Penno.
14. Ellesiyeh, Eighteenth Dynasty Chapel, Thutmose III.
15. Derr, Nineteenth Dynasty Temple, Ramses II.
16. Amada, Eighteenth Dynasty Temple, Thutmose III, Amenhotep II, Thutmose IV.
17. Gerf Husên, Nineteenth Dynasty Temple, Ramses II.
18. Bêt el-Wâli, Nineteenth Dynasty Temple, Ramses II.

www.ingramcontent.com/pod-product-compliance
Lightning Source LLC
Chambersburg PA
CBHW041927260326
41914CB00009B/1205